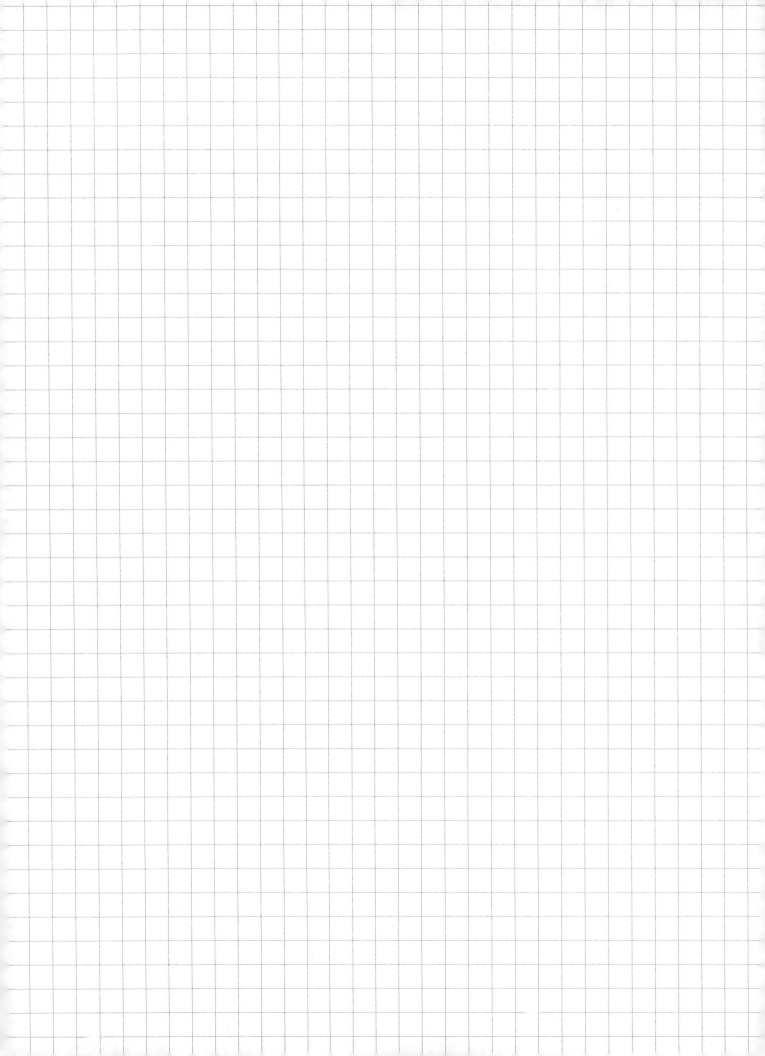

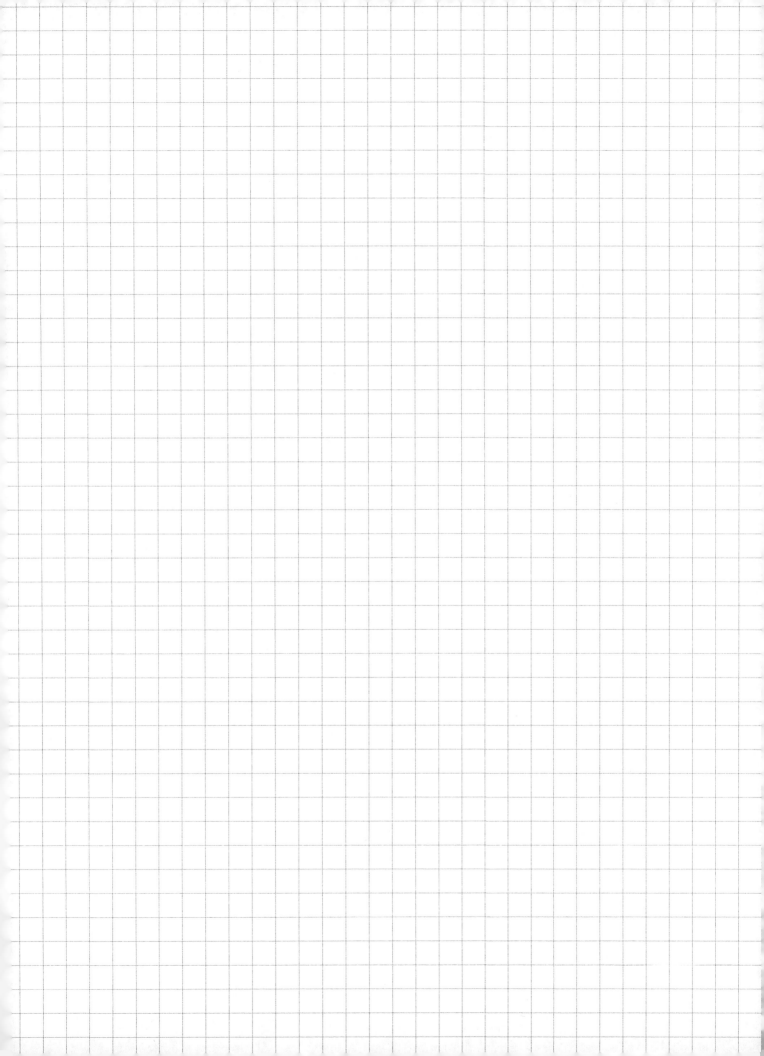

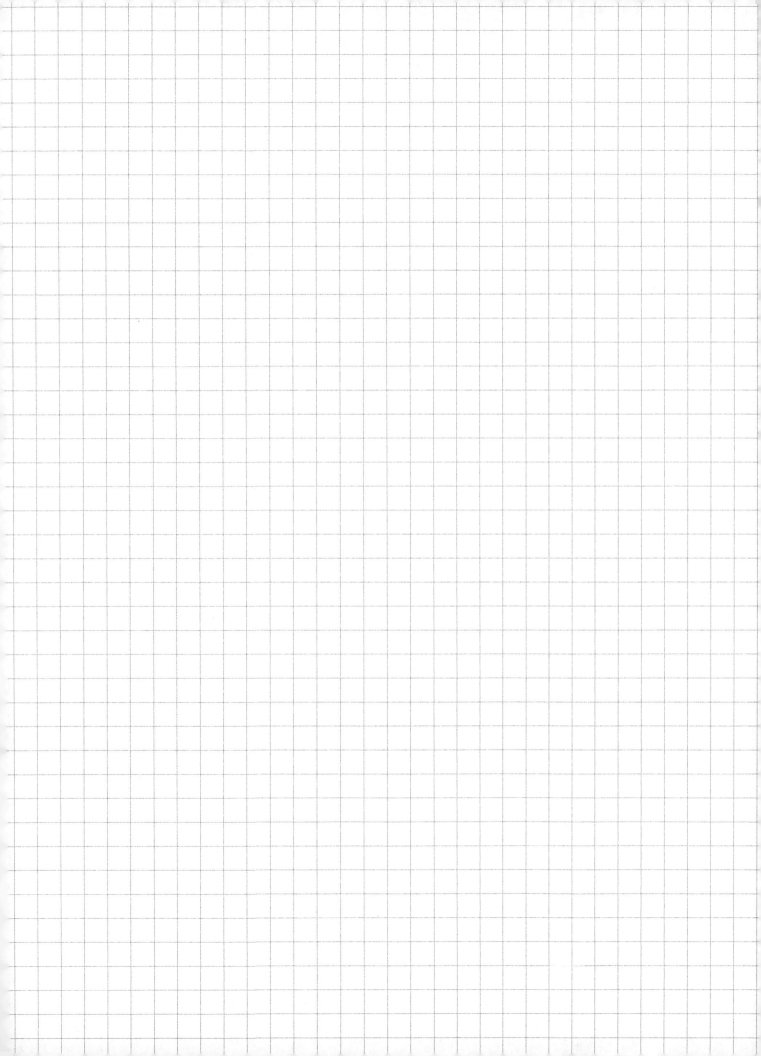

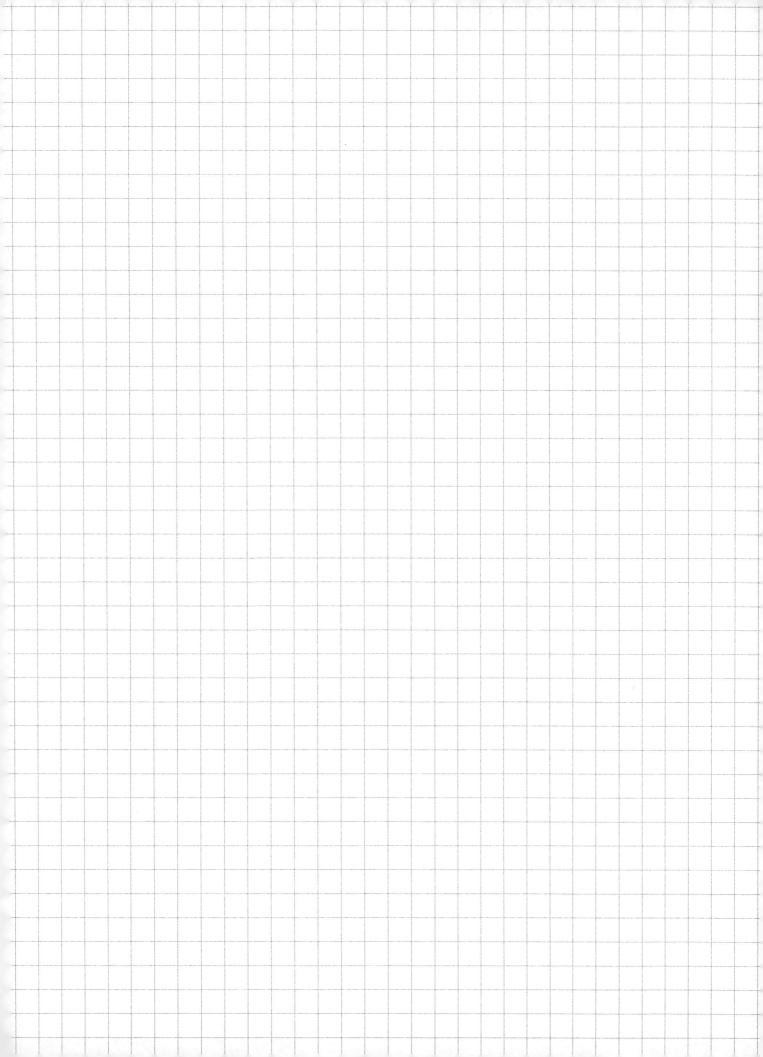

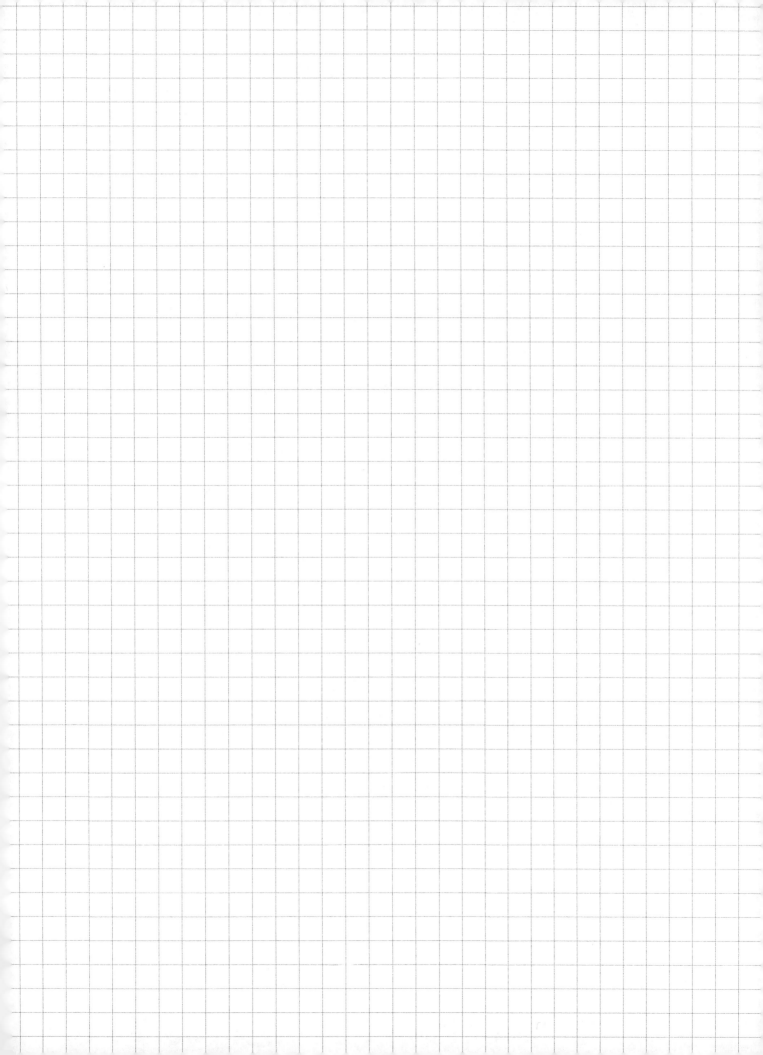

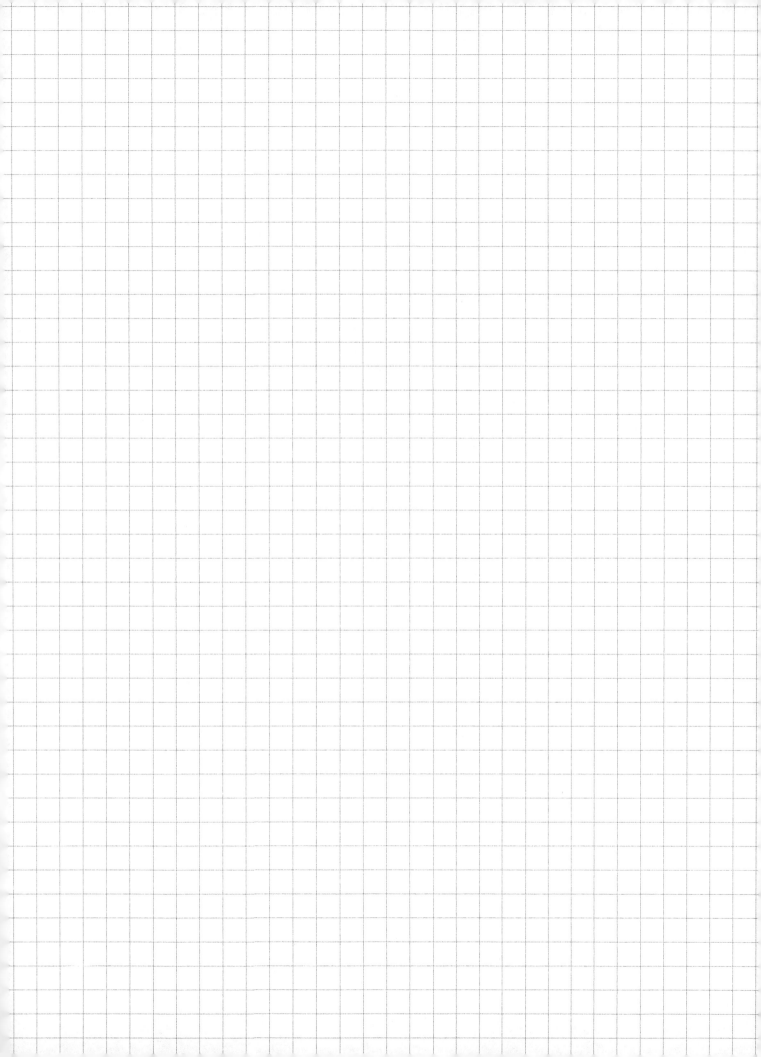

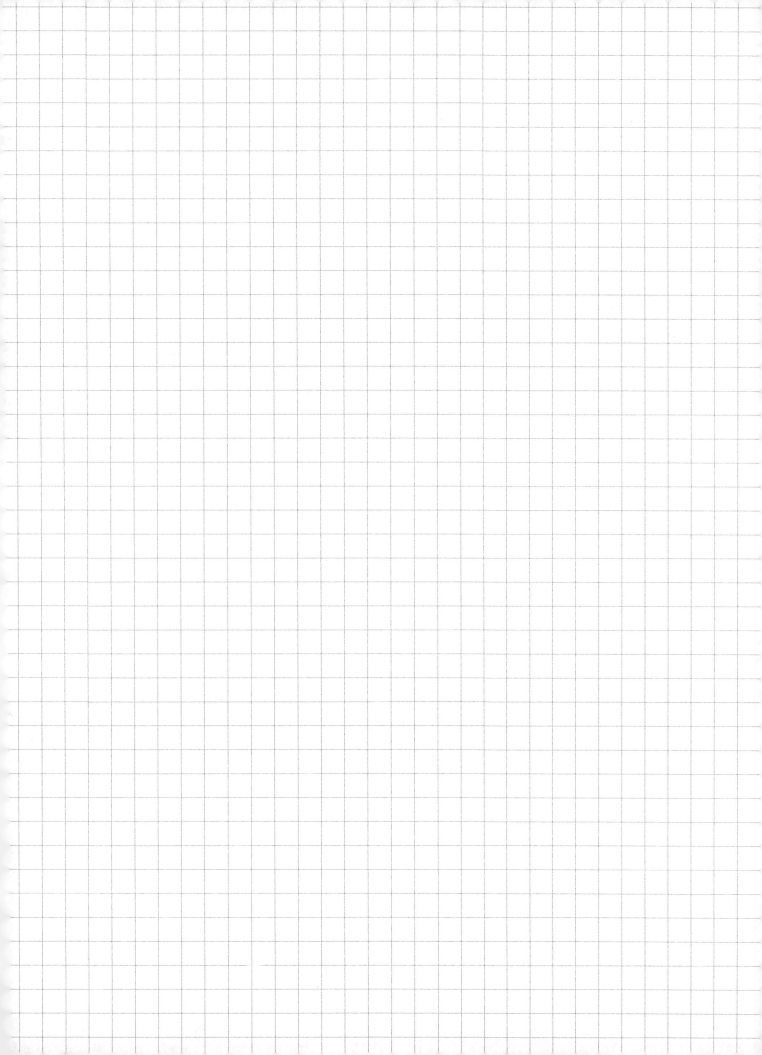

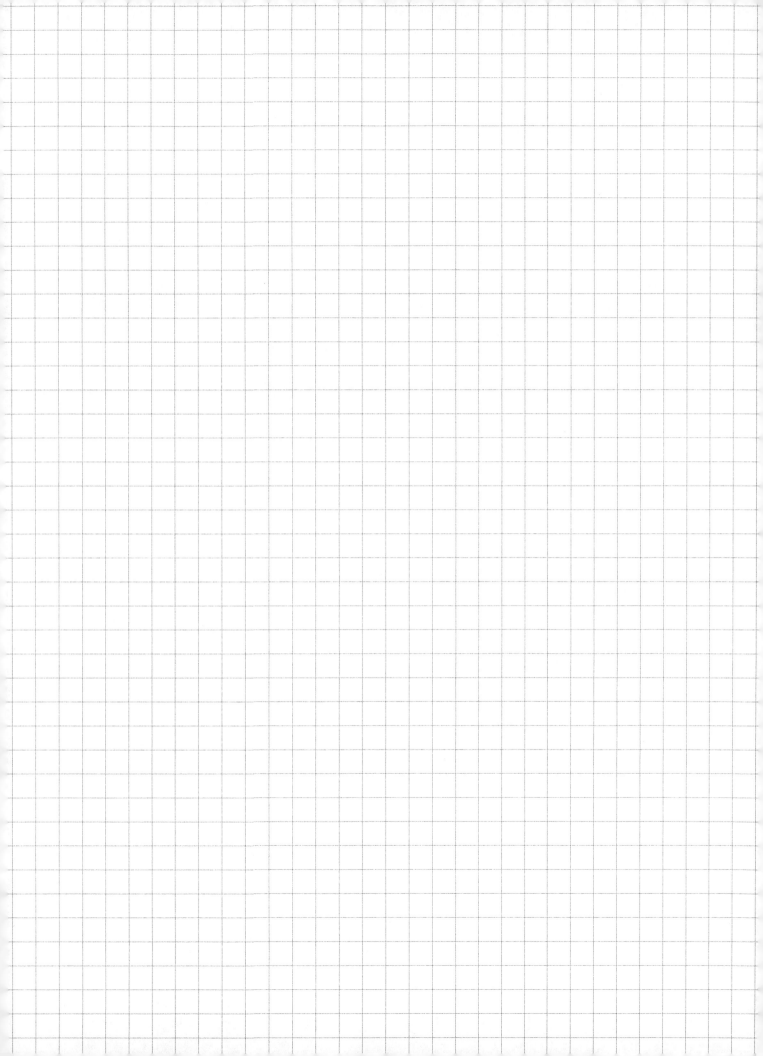

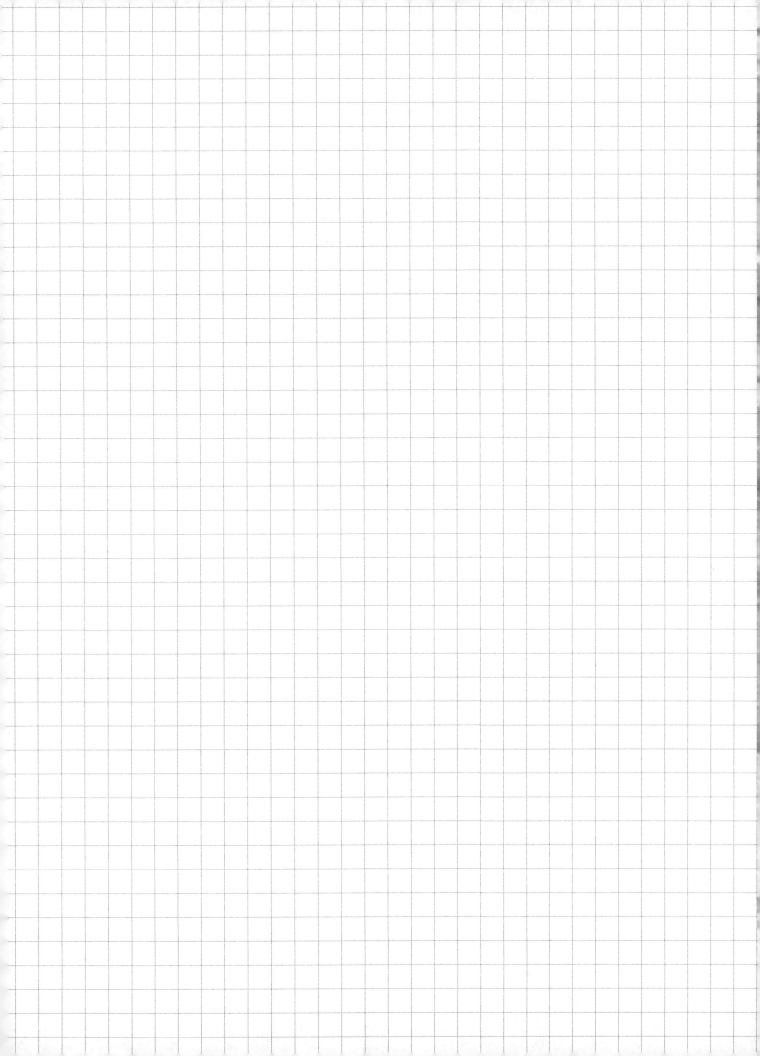

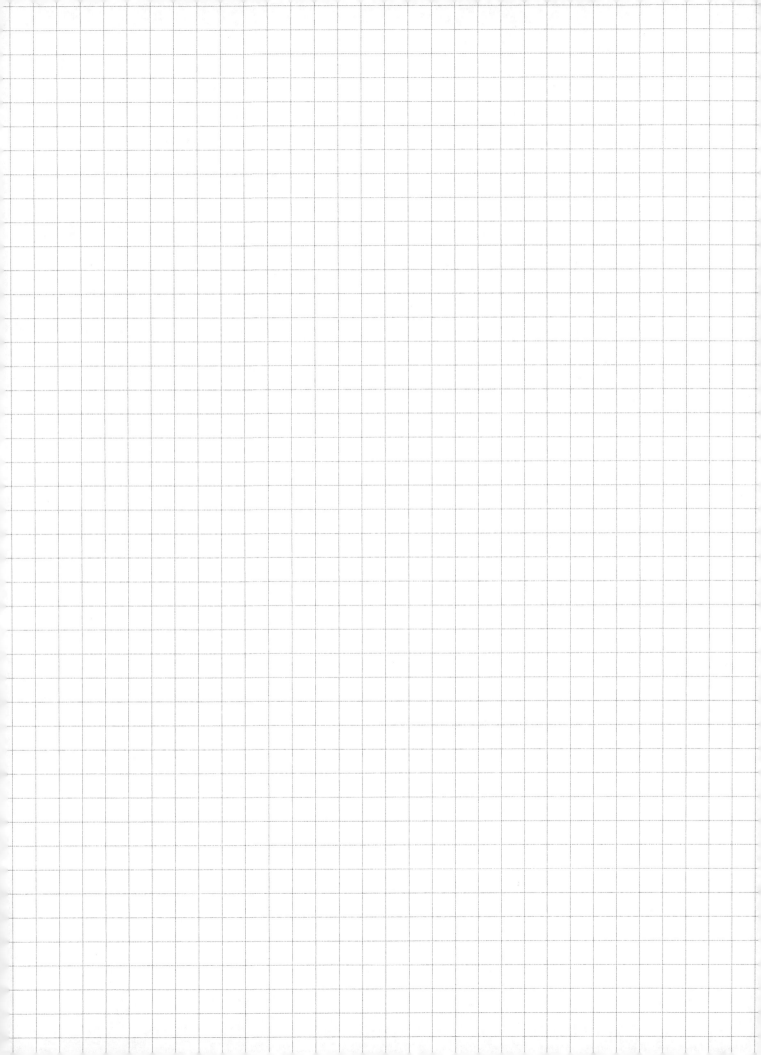

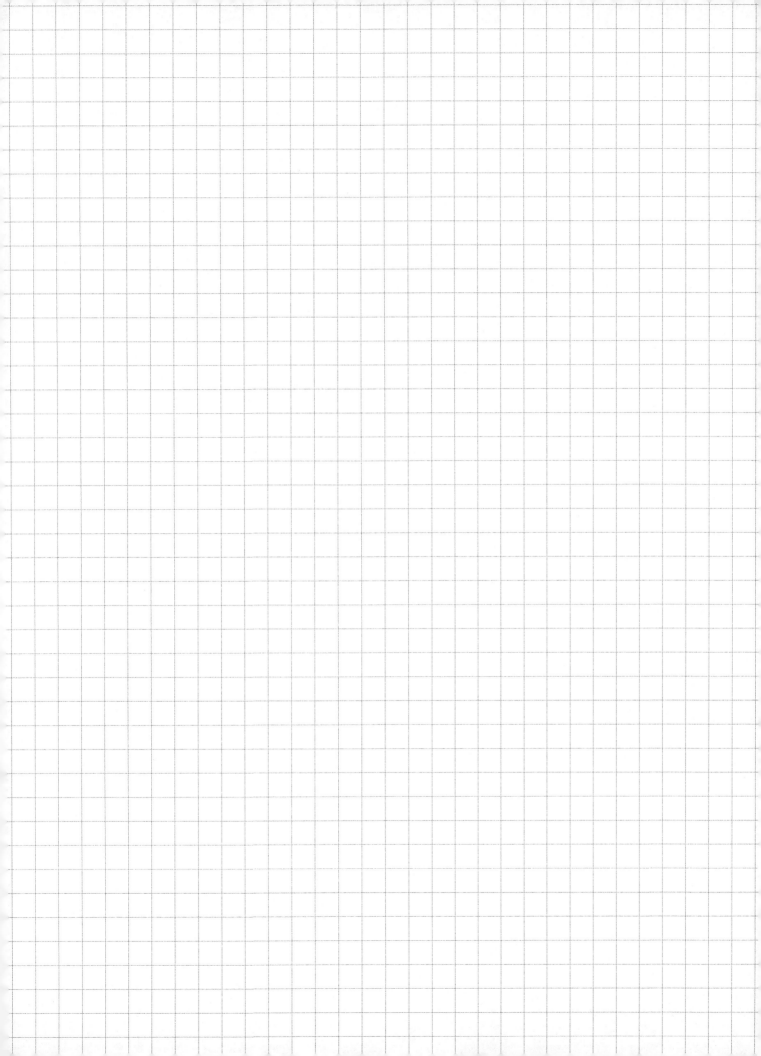

Made in the USA
Las Vegas, NV
16 August 2024

93902602R00063